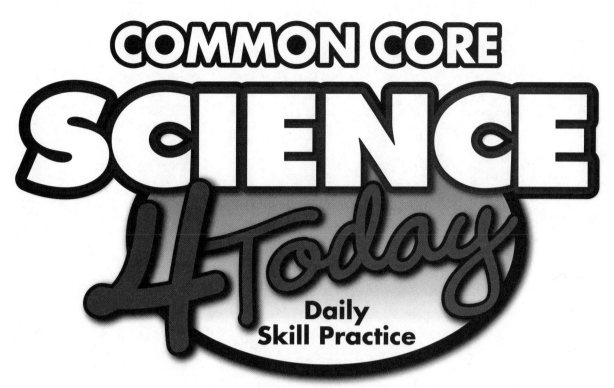

COMMON CORE SCIENCE 4 Today

Daily Skill Practice

Kindergarten

Jennifer B. Stith

Carson-Dellosa Publishing, LLC
Greensboro, North Carolina

Credits

Content Editor: Elise Craver
Copy Editor: Karen Seberg

 Visit *carsondellosa.com* for correlations to Common Core, state, national, and Canadian provincial standards.

Carson-Dellosa Publishing, LLC
PO Box 35665
Greensboro, NC 27425 USA
carsondellosa.com

ISBN 978-1-4838-1123-9
01-135141151

Table of Contents

Common Core Science 4 Today is a perfect supplement to any classroom science curriculum. Students' science skills will grow as they support their knowledge of science topics with a variety of engaging activities.

This book covers 40 weeks of daily practice. You may choose to work on the topics in the order presented or pick the topic that best reinforces your science curriculum for that week. During the course of four days, students take about 10 minutes to complete questions and activities focused on a science topic. On the fifth day, students complete a short assessment on the topic.

Various skills and concepts in math and English language arts are reinforced throughout the book through activities that align to the Common Core State Standards. Due to the nature of the Speaking and Listening standards, classroom time constraints, and the format of the book, students may be asked to record verbal responses. You may wish to have students share their answers as time allows. To view these standards, please see the Common Core State Standards Alignment Matrix on pages 5–8.

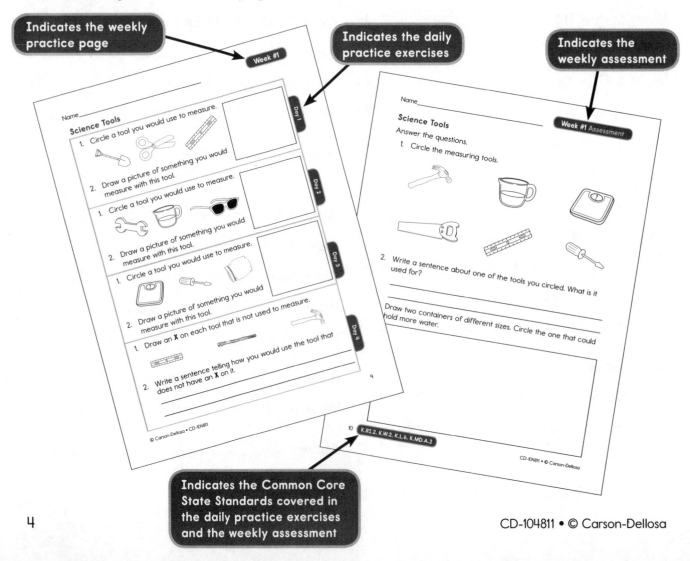

Indicates the weekly practice page

Indicates the daily practice exercises

Indicates the weekly assessment

Indicates the Common Core State Standards covered in the daily practice exercises and the weekly assessment

CD-104811 • © Carson-Dellosa

English Language Arts

STANDARD	W1	W2	W3	W4	W5	W6	W7	W8	W9	W10	W11	W12	W13	W14	W15	W16	W17	W18	W19	W20
K.RI.1		●				●	●					●								
K.RI.2	●																			
K.RI.3													●						●	
K.RI.4																				●
K.RI.5																				
K.RI.6																				
K.RI.7		●	●	●	●	●	●	●	●	●	●	●	●	●	●	●	●	●	●	●
K.RI.8																				
K.RI.9																				
K.RI.10																				
K.RF.1																				
K.RF.2																				
K.RF.3																				
K.RF.4						●		●	●	●	●	●	●	●	●	●	●	●	●	●
K.W.1																				
K.W.2	●	●	●	●	●	●		●	●		●			●	●	●			●	
K.W.3																				
K.W.5																				
K.W.6																				
K.W.7																				
K.W.8																				
K.SL.1			●	●	●	●	●							●					●	
K.SL.2													●							
K.SL.3																				
K.SL.4																				
K.SL.5																				
K.SL.6								●												
K.L.1																				
K.L.2																				
K.L.3																				
K.L.4				●	●				●	●				●		●			●	●
K.L.5		●	●								●				●			●		
K.L.6	●					●	●	●												

W = Week

English Language Arts

STANDARD	W21	W22	W23	W24	W25	W26	W27	W28	W29	W30	W31	W32	W33	W34	W35	W36	W37	W38	W39	W40
K.RI.1		●								●	●		●		●	●	●			●
K.RI.2																				
K.RI.3																				
K.RI.4	●		●	●	●				●							●				
K.RI.5																				
K.RI.6																				
K.RI.7	●	●	●	●	●	●	●	●	●	●	●	●	●	●	●	●	●	●	●	●
K.RI.8																				
K.RI.9																				
K.RI.10																				
K.RF.1																				
K.RF.2																				
K.RF.3																				
K.RF.4	●	●	●	●	●	●	●	●	●	●	●	●	●	●	●	●	●	●	●	●
K.W.1																				
K.W.2		●			●	●			●	●		●	●	●				●	●	
K.W.3																				
K.W.5																				
K.W.6																				
K.W.7																				
K.W.8																				
K.SL.1						●	●	●							●		●			●
K.SL.2													●						●	
K.SL.3																				
K.SL.4																				
K.SL.5																				
K.SL.6																				
K.L.1																				
K.L.2																				
K.L.3																				
K.L.4	●	●	●	●	●	●	●	●	●	●	●	●	●	●	●	●	●	●	●	●
K.L.5														●						
K.L.6																				

W = Week

CD-104811 • © Carson-Dellosa

Math

STANDARD	W1	W2	W3	W4	W5	W6	W7	W8	W9	W10	W11	W12	W13	W14	W15	W16	W17	W18	W19	W20
K.CC.A.1																				
K.CC.A.2																				
K.CC.A.3																	●	●		●
K.CC.B.4																				
K.CC.B.5							●									●		●		
K.CC.C.6							●					●	●							
K.CC.C.7											●								●	●
K.OA.A.1										●										
K.OA.A.2													●	●						
K.OA.A.3																				
K.OA.A.4										●										
K.OA.A.5												●								
K.NBT.A.1																				
K.MD.A.1											●									
K.MD.A.2	●		●			●			●		●									
K.MD.B.3			●				●										●	●		●
K.G.A.1				●																
K.G.A.2									●											
K.G.A.3																				
K.G.B.4			●																	
K.G.B.5				●																
K.G.B.6																				

W = Week

Common Core State Standards Alignment Matrix

Math

STANDARD	W21	W22	W23	W24	W25	W26	W27	W28	W29	W30	W31	W32	W33	W34	W35	W36	W37	W38	W39	W40
K.CC.A.1																				
K.CC.A.2																				
K.CC.A.3	●		●			●		●	●		●			●						
K.CC.B.4							●								●					
K.CC.B.5								●						●	●					
K.CC.C.6												●								
K.CC.C.7																				
K.OA.A.1												●								
K.OA.A.2					●	●										●	●			
K.OA.A.3																				
K.OA.A.4												●								
K.OA.A.5																				
K.NBT.A.1								●												
K.MD.A.1																				
K.MD.A.2																				
K.MD.B.3												●								
K.G.A.1																				
K.G.A.2																				
K.G.A.3																				
K.G.B.4																				
K.G.B.5																				
K.G.B.6																				

W = Week

CD-104811 • © Carson-Dellosa

Science Tools

1. Circle a tool you would use to measure.

 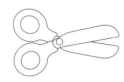 

2. Draw a picture of something you would measure with this tool.

---

1. Circle a tool you would use to measure.

2. Draw a picture of something you would measure with this tool.

---

1. Circle a tool you would use to measure.

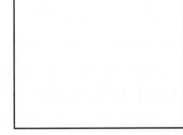

2. Draw a picture of something you would measure with this tool.

---

1. Draw an **X** on each tool that is not used to measure.

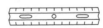  

2. Write a sentence telling how you would use the tool that does not have an **X** on it.

_____

_____

Science Tools

Answer the questions.

1. Circle the measuring tools.

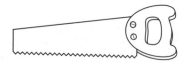 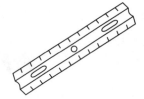 

2. Write a sentence about one of the tools you circled. What is it used for?

   _____

   _____

3. Draw two containers of different sizes. Circle the one that could hold more water.

The Five Senses

Complete each sentence. Use the words from the word bank.

ears   hand   mouth

1. I use my _____ to hear birds sing.

2. I use my _____ to touch a soft kitten.

3. I use my _____ to taste an apple.

1. Circle the objects you can smell.

2. Circle the objects you can see.

1. Draw a picture of something you can see and taste.

2. Draw a picture of something you can touch and smell.

Draw a line to connect each body part with the object it senses.

1.   2.   3.

The Five Senses

Follow the directions to complete the picture.

1. Draw with red the parts you use to hear.

2. Draw with green the parts you use to touch.

3. Draw with blue the part you use to smell.

4. Draw with orange the parts you use to see.

5. Draw with purple the part you use to taste.

K.RI.1, K.RI.7, K.W.2, K.L.5

Classifying

1. Color the objects that are alike.

2. Tell or write how they are alike.

_____

1. Circle the objects that are alike.

2. Tell or write how they are alike.

_____

1. Circle the objects that are alike.

2. Tell or write how they are alike.

_____

1. Circle the objects that are alike.

 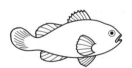 

2. Tell or write how they are alike.

_____

Day 1

Day 2

Day 3

Day 4

Classifying

Answer the questions.

Draw an **X** on the object in each row that is different.

1.

2.

3.

4.

5.

6. Write a sentence about one of the rows above. Tell why the object with the **X** does not belong.

_____

_____

K.RI.7, K.W.2, K.SL.1, K.L.5, K.MD.A.2, K.MD.B.3, K.G.B.4

Position

1. Circle the correct word or phrase to complete the sentence.

   The circle is (on top of, behind) the square.

2. Draw a triangle on top of a square.

1. Circle the picture that matches the sentence.

2. Talk with a friend about where the circle is.

The dot is beside the rectangle.

1. Draw a dog in front of the doghouse.

2. Draw a bone beside the dog.

1. Write a sentence about where the apple is.

   _____

   _____

   _____

   _____

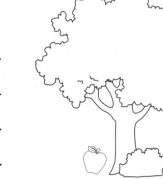

Position

Read each sentence. Draw a line to match each sentence to the correct picture.

1. The sun is above the cloud.

2. The sun is behind the cloud.

3. The ball is on top of the box.

4. The ball is below the box.

Follow the directions.

5. Draw a car beside the truck.
6. Draw a box on top of the truck.
7. Draw a building behind the truck.

K.RI.7, K.W.2, K.SL.1, K.L.4, K.G.A.1, K.G.B.5   CD-104811 • © Carson-Dellosa

Types of Motion

1. Circle the words that describe how things move.

kick	zigzag	up and down
walk	straight	round and round
back and forth	lift	fast and slow

1. Draw something that moves up and down.

2. Write a sentence about the object you drew.

_____

_____

1. Circle the objects that go round and round.

2. Write the name of another object that goes round and round.

_____

Complete the sentences.

1. An elevator moves _____.

2. A wheel on a bike goes _____.

3. A rolled ball moves in a _____ line.

Types of Motion

Answer the questions.

1. Draw arrows to show how each object moves.

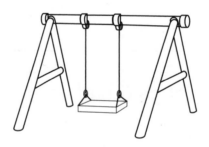

2. Move in one of the ways below. Ask a friend to tell how you moved.

    straight   round and round   zigzag

        back and forth   up and down

How Things Move

1. Color the things that move by themselves. Circle the one that is slow.

1. Circle the things that do not move by themselves.

2. Write a sentence telling how to make a ball move fast.

_____

1. Draw an **X** on the object that does not belong.

2. Talk with a friend about why the object with the **X** does not belong.

Read each question. Circle **yes** or **no**.

1. Is a car faster than a bike? yes no

2. Is a turtle faster than a tiger? yes no

3. Is it faster to run than to walk? yes no

Name_____

How Things Move

Answer the questions.

1. Draw two things that move fast.

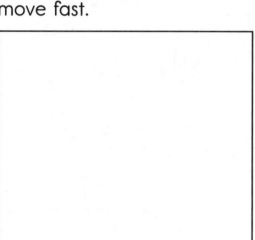

2. Draw two things that move slowly.

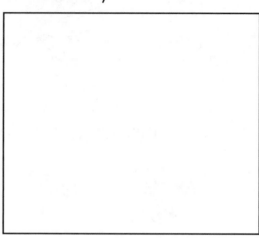

3. Circle each sentence that is **true**.

   A race car moves slowly.

   A horse runs slowly.

   A top spins fast.

   A fan moves by itself.

4. Circle the ramp where a marble would roll down faster. Draw an **X** on the ramp where a marble would roll down slower. Talk with a friend or write a sentence telling why you think so.

_____

_____

K.RI.1, K.RI.7, K.RF.4, K.W.2, K.SL.1, K.L.6, K.MD.A.2

Name_____

Forces

Complete each sentence. Use the words from the word bank.

force	pull	push

1. A push or pull is a _____.

2. She will _____ open the oven door.

3. He will _____ the buttons on the phone.

1. Circle the word that describes the picture.

    push   pull

2. Color the object being pulled in the picture.

1. Circle the word that describes the picture.

    push   pull

2. Color the object being pushed in the picture.

1. Act out a push or pull. Draw a picture of what you did.

Forces

Answer the questions.

1. Look at the objects. Write the force needed to make the objects in each group move.

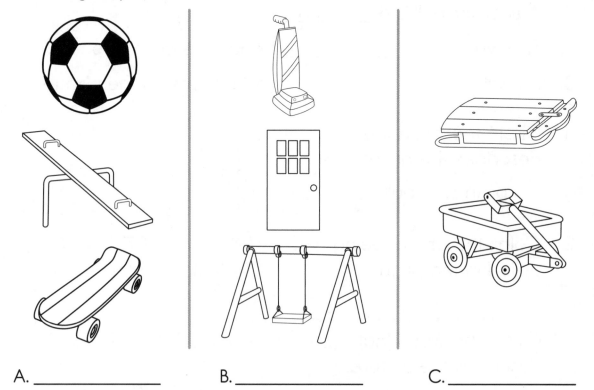

   A. _____   B. _____   C. _____

2. How many objects are in each group?

    A. _____   B. _____   C. _____

3. Which group has the fewest objects? _____

4. How are the objects sorted?

    _____

5. Talk to a friend about your answers to 2 to 4.

Name_____

Gravity

1. Circle the picture that shows something falling to the ground.

1. Complete the sentence. Use the words from the word bank.

| down | gravity |

_____is the force that pulls a

child _____ a slide.

1. Tell how both pictures show gravity.

_____

_____

_____

_____

_____

1. Look around the room. Write the names of four objects being held down by gravity.

_____   _____

_____   _____

Gravity

Complete each sentence. Use the words from the word bank.

| falling | floating | force | gravity |

1. Gravity is a _____.

2. Gravity pulls _____ objects to the ground.

3. Gravity keeps objects from _____ away.

4. The apple falls to the ground because _____ pulls it down.

K.RI.7, K.RF.4, K.W.2, K.SL.6, K.L.6   CD-104811 • © Carson-Dellosa

Size and Shape

1. Color the animal that is largest.

2. Color the fruit that is smallest.

1. Color the pizza that is different.

2. Write a sentence telling how it is different.

_____

_____

1. Circle the word that tells how these leaves are different.

size   shape

1. Draw a picture of two houses. Make sure they are different in size and shape.

Size and Shape

Look at the sizes and shapes of the shoes. Color each matching pair of shoes the same color.

Color and Texture

1. Circle the color words.

   red   brown

   striped   green

   smooth   black

2. Draw a picture of something that is green.

Color the objects in each row that are the same color.

1.

2.

1. Circle the words that tell about texture, or how something feels.

   blue   soft

   hard   white

   smooth   rough

2. Draw a picture of something that is soft.

Circle the color and texture words in the sentences.

1. The white bunny is soft.

2. The brown chair is hard.

3. The pink shell is smooth.

Color and Texture

Answer the questions.

1. Draw lines to match the objects to the word or words that describe them.

               soft

               hard

               rough

               smooth

               black

               yellow

2. Sam has 10 marbles in all. He has 3 green marbles. The rest are blue. Draw a picture of Sam's marbles.

   How many marbles are blue? _____

K.RI.7, K.RF.4, K.L.4, K.OA.A.1, K.OA.A.4

Weight

1. Circle the objects that are light. Draw an **X** on each object that is heavy.

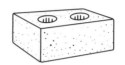

1. Order the objects **1** to **3** from lightest to heaviest.

_____   _____   _____

1. Circle each sentence that is **true**.

A baby is lighter than a house.

A truck is heavier than a book.

A shoe is heavier than a tree.

1. Draw a picture of an object in the room that is light.

2. Draw a picture of an object in the room that is heavy.

Weight

Answer the questions.

1. Draw a line to match each object to the word that describes its weight.

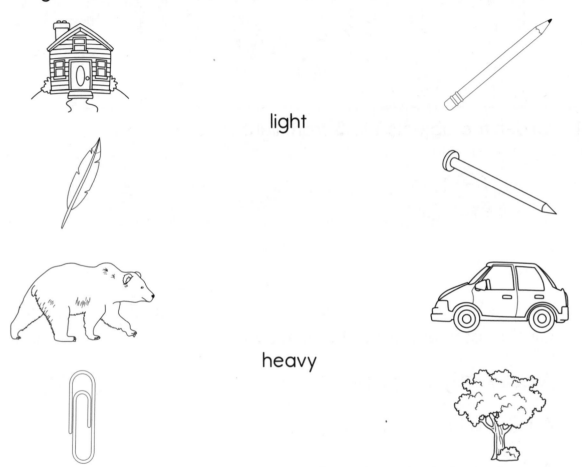

light

heavy

2. Jan weighed the coin and the pencil. The pencil was heavier than the coin. Circle the correct number sentence.

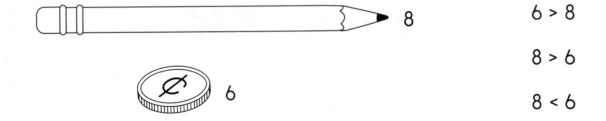

6 > 8

8 > 6

8 < 6

Float or Sink

Day 1

1. Draw a picture of yourself in a pool.

2. Do you float or sink?

_____

Day 2

1. Color the objects that sink.

2. Color the objects that float.

Day 3

1. Circle each sentence that is **true**.

A hammer will float.

A straw will float.

A feather will sink.

A penny will float.

Day 4

1. Look around the room. Write the name of one object you think will float. Write the name of one object you think will sink. Talk with a friend about why you think so.

float _____

sink _____

Float or Sink

Answer the questions.

1. Look at each object. Decide if it will float or sink. Draw an **X** in the correct box.

	float	sink
A.		
B.		
C.		
D.		
E.		

2. How many objects will float? _____

3. How many objects will sink? _____

4. How many objects are there in all? _____

Materials

1. Look at the first object in the row. Circle the object that is made of the same material.

1. Look at the first object in the row. Circle the object that is made of the same material.

1. Draw a line to match each object to the material it is made from.

wood   metal

1. Draw a picture of an object in the room that is made of both wood and metal.

Materials

Answer the questions.

1. Color the objects made of wood brown. Color the objects made of metal blue.

 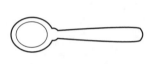 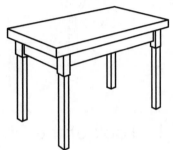

2. Circle the name of the set that has more objects.

   wood   metal

3. How many more metal objects than wood objects are there?

   _____

Materials and Their Uses

1. Circle the furniture.

2. What material is each made of? _____

Day 1

1. Draw a picture of a metal object you cook with.

Day 2

1. Circle the name of each object you can wear.

   pants   can   sock   pan

   coat   hat   pen

2. Talk to a friend about what materials you think are used to make clothes and why.

Day 3

1. Choose one of the objects below. Write a sentence telling why you would not want it.

   a pan made of wood   a pillow made of metal

   a saw made of cloth

   _____

   _____

Day 4

Materials and Their Uses

Answer the questions.

1. Circle each material that can be used to make a table.

   wood   metal   cloth

2. Write a sentence telling why a table is not made of cloth.

   _____

   _____

3. Circle each material that can be used to make clothes.

   wood   metal   cloth

4. Write a sentence telling why clothes are not made of wood or metal.

   _____

   _____

5. Two trees were cut down to make tables. The wood from one tree made 5 tables. The wood from the other tree made 3 tables. How many tables were made in all? Draw a picture to help you solve the problem.

   _____

Physical Change

1. Circle each picture that shows a physical change.

1. Circle each action that causes a physical change.

stir	cut	burn
break	fold	melt

1. Write a sentence telling why the picture shows a physical change.

    _____

    _____

    _____

    _____

1. Choose one of the pictures below. Act out the change. Draw a picture of the change.

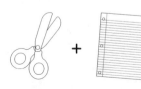  

Physical Change

Answer the questions.

1. Circle each physical change.

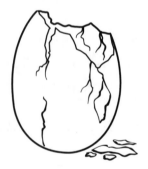

2. Choose one picture. Write a sentence telling what is happening.

_____

_____

Water

Complete each sentence. Use the words from the word bank.

liquid	poured	shape

1. Water is a _____.

2. It takes the _____ of its container.

3. Water can be _____.

1. Circle the sentence that is **true**.

   When you put water in a cold freezer, it turns into a gas.

   When you put water in a cold freezer, it turns into a solid.

1. Draw a picture showing what happens to ice when you take it out of a freezer.

1. Write a sentence telling what is happening to the water in the pot.

   _____

   _____

   _____

   _____

Water

Answer the questions.

1. Draw a picture to match each sentence.

Water freezes and turns into ice.

Ice melts and turns into water.

Water boils and turns into steam.

2. How many states of matter can water be? _____

3. Draw a line to match each form of water to the correct state of matter.

water   solid

ice   liquid

steam   gas

Name_____

Living and Nonliving Things

1. Circle the living things.

2. Circle the nonliving things.

1. Draw a picture of a living thing you find outside.

2. Draw a picture of a nonliving thing you find in a room.

1. Write two things that an animal needs to live.

    _____   _____

2. A rock is a nonliving thing. Circle each thing a rock cannot do.

    eat   breathe   sleep

    roll   drink   break

Choose the best word to complete each sentence.

1. A caterpillar is alive. It _____ leaves.

    smells   eats

2. A tiger is alive. It _____ for food.

    hunts   sleeps

Living and Nonliving Things

Answer the questions.

1. Count the objects in each group. Write the total number. Draw a line to match each group to its name.

   A. _____

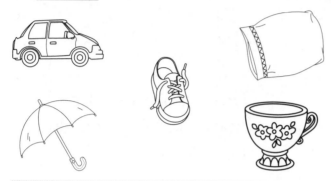

living things

   B. _____

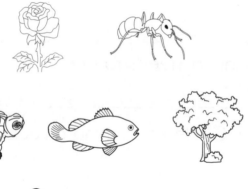

nonliving things

2. Circle with red each word that describes what animals do.
   Circle with blue each word that describes what plants do.
   You may circle some words twice.

eat food	make food	breathe
move	drink	grow

Name_____

Classifying Plants and Animals

1. Draw an **X** on the animal that does not belong.

2. Write what the three animals have in common.

_____

1. Draw an **X** on the plant that does not belong.

2. Write what the three plants have in common.

_____

1. Draw an **X** on the animal that does not belong.

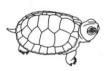  

2. Write what the three animals have in common.

_____

1. Draw an **X** on the animal that does not belong.

2. Write what the three animals have in common.

_____

Classifying Plants and Animals

Answer the questions.

1. Circle each animal. Draw an **X** on each plant.

2. How many animals are there? _____

3. How many plants are there? _____

4. Which group has more? _____

 K.RI.7, K.RF.4, K.L.4, K.L.5, K.CC.A.3, K.CC.B.5, K.MD.B.3   CD-104811 • © Carson-Dellosa

Name_____

Animal Offspring

1. Circle this mother's baby.

1. Draw an X on the name of each animal that is not a dog's baby.

    cub   joey   kitten

    kid   puppy   bunny

1. Write a sentence telling why these two animals are not mother and baby.

_____

_____

_____

_____

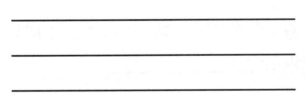 

1. Draw your favorite animal and the animal's baby. Talk with a friend about your drawings.

Animal Offspring

Answer the questions.

1. Draw a line to match each adult name to the baby name.

    cow   foal

    duck   duckling

    cat   chick

    horse   calf

    frog   tadpole

    chicken   kitten

2. A mother duck has 6 ducklings. A mother dog has 9 puppies.
    Which mother has more babies? _____
    Write a comparison sentence using numbers and >, <, or =.

K.RI.3, K.RI.7, K.RF.4, K.W.2, K.SL.1, K.L.4, K.CC.C.7   CD-104811 • © Carson-Dellosa

Name_____

Comparing Animals

1. Circle the names of the animals that have fur.

    lion   fish   owl

    cat   snake   horse

1. Color the animals that have feathers.

1. Draw scales on the animals that need them.

 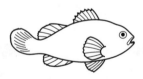  

1. Write the word that names each animal's covering. Use the words from the word bank.

feathers	fur	scales

_____   _____   _____

Comparing Animals

Answer the questions.

1. Circle each animal that has fur.

   Draw an **X** on each animal that has scales.

   Color each animal that has feathers.

2. Count the number of animals with each type of covering. Write the numbers.

   fur _____   scales _____   feathers _____

3. Compare how many are in each group. Circle the word you used to compare.

   more   less   equal

K.RI.4, K.RI.7, K.RF.4, K.L.4, K.CC.A.3, K.CC.C.7, K.MD.B.3   CD-104811 • © Carson-Dellosa

Parts of the Body

Complete each sentence with a number.

1. Your face has _____ eyes.

2. You have _____ ears on your head.

3. You have _____ nose.

1. Circle the sentences that are **true**.

    Legs and feet help you walk.

    Ears help you eat.

    Fingers help you tie your shoes.

1. Draw a picture of the body part you use to eat.

2. Draw a picture of the body part you use to kick.

1. Circle the body parts that help you see.

2. Circle the body part that helps you open a door.

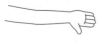

Parts of the Body

Answer the question.

1. Write the parts of the body. Use the words from the word bank.

| elbow | eye | foot | hand | knee | nose |

K.RI.4, K.RI.7, K.RF.4, K.L.4, K.CC.A.3   CD-104811 • © Carson-Dellosa

Animal Parts and Their Functions

1. Some animals that live in water have fins. Circle the words that tell how fins help animals.

   help them eat   help them move   help them see

1. Many animals have tails. Draw a picture of an animal using its tail.

1. Circle each sentence that is **true**.

   Fur helps keep an animal dry.

   Fur helps keep an animal warm.

1. A rabbit has large ears so that it can hear sounds from far away. Circle the other animals that can hear faraway sounds.

Animal Parts and Their Functions

Answer the questions.

Read each riddle. Write the name of the animal part. Use the words from the word bank.

ears	fins	hands	tail	wings

1. A monkey uses this to swing from trees. _____

2. A hawk uses these to fly. _____

3. A deer uses these to hear faraway sounds. _____

4. A shark uses these to swim fast in water. _____

5. A person uses these to get dressed. _____

6. Draw a picture of your favorite animal. Write one sentence telling how it uses one of its body parts.

_____

_____

_____

_____

_____

K.RI.1, K.RI.7, K.RF.4, K.W.2, K.L.4

Name_____

Parts of a Plant

1. Circle the words that name plant parts. Write the number of words you circled. _____

| stem | knee | toes | flower |

| leaf | nose | roots | elbow |

1. Circle the sentences that are **true**.

Roots are above the ground.

Roots are under the ground.

Most leaves are green.

Most leaves are pink.

1. Circle the plant parts.

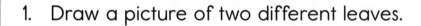

1. Draw a picture of two different leaves.

Parts of a Plant

Color the parts of the plant.

1. Color the flower red.

2. Color the stem yellow.

3. Color the leaves green.

4. Color the roots brown.

K.RI.4, K.RI.7, K.RF.4, K.L.4, K.CC.A.3

Plant Parts and Their Functions

1. Circle the best words to complete the sentences.

   Many plants have colorful (flowers, roots). The bright colors attract bees and other insects. A (stem, fruit) grows from a flower.

1. Circle the sentence that is **true**.

   Leaves help a plant make food from the sun.

   Leaves help a plant stay dry in the rain.

1. Roots hold a plant in place. Roots take in water for the plant. Draw roots on this plant.

1. A plant's seeds are inside the fruit. Color the foods that have seeds.

Plant Parts and Their Functions

Complete each sentence.

1. The _____ makes food for the plant.

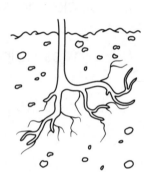

2. The _____ hold the plant in soil. They take in water for the plant.

3. The _____ makes seeds for the plant.

4. The _____ holds the leaves and flowers. It carries water and food to all parts of the plant.

Week #25

Habitats

1. Circle the word that completes the sentence.

   A _____ is the natural home of a plant or animal.

   zoo   habitat   house

1. Circle the names of animals that live in this habitat.

   owl   goat

   bear   fox

1. Draw a habitat for these animals.

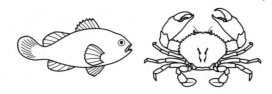

1. Write a sentence about a spider's habitat.

   _____

   _____

Habitats

Answer the questions.

1. Draw a line to match each animal to its habitat.

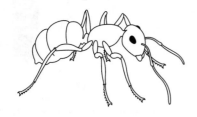
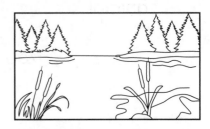

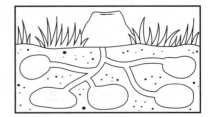

2. Underground are 7 ants. Also underground are 3 worms. How many animals are underground in all? _____

K.RI.4, K.RI.7, K.RF.4, K.W.2, K.L.4, K.OA.A.2

Desert Habitat

1. Choose the best word to complete the sentence.

   A desert is a place that gets little rain. It is _____.

   wet   dry   hot

Day 1

1. Circle the names of the animals that could live in this home.

   lizard   kangaroo

   hawk   mouse

Day 2

1. Draw a home for this animal.

2. Talk with a friend about why the home is good for a snake.

Day 3

1. Write a sentence telling why this plant might not be a good home for some animals.

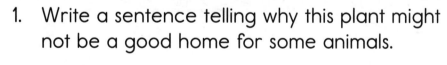

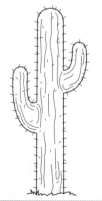

Day 4

Desert Habitat

Answer the questions.

1. Look at the animals in the desert. Draw an **X** on each animal that does not belong.

2. Draw 2 big rocks in the picture. Draw 4 small rocks in the picture. Write the total number of rocks you drew. _____

 K.RI.7, K.RF.4, K.W.2, K.SL.1, K.L.4, K.CC.A.3, K.OA.A.2 CD-104811 • © Carson-Dellosa

Name_____

Polar Habitat

Circle the best word to complete each sentence.

1. A polar habitat is very (warm, cold).

2. A (panda, polar bear) lives in a polar habitat.

1. Circle the name of the animal that lives in a polar habitat.

   ant   mole   penguin   snake

2. Write the name of another animal that lives in a polar habitat.

   _____

1. Draw a silly home for this animal. Talk with a friend about why it cannot live there.

1. Write a sentence about the color of some polar animals' fur.

   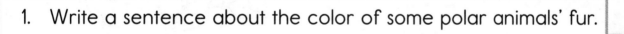

   _____

   _____

Polar Habitat

Answer the questions.

1. Draw a polar habitat for these animals.

2. Draw 2 more seals in the picture. How many polar animals are there in all? _____

Ocean Habitat

1. Choose the best word to complete the sentence.

   An ocean is made up of _____.

   water   rock   air

1. Draw a picture of an animal living in an ocean habitat.

1. Act out what it is like to live underwater. Have a friend describe what you are doing.

2. Write the words your friend used.

   _____

   _____

1. Circle the names of the animals that live in water.

   whale   mouse   shark   crab

   cat   fish   horse   dolphin

Ocean Habitat

Answer the questions.

1. Count the sharks. Write the number. _____

2. Count the seahorses. Write the number. _____

3. Count the crabs. Write the number. _____

4. Count the fish. Circle 10 fish. Write the total number of fish. _____

K.RI.7, K.RF.4, K.SL.1, K.L.4, K.CC.A.3, K.CC.B.5, K.NBT.A.1

Objects in the Sky

Complete each sentence. Use the words from the word bank.

moon	stars	sun

1. The _____ gives Earth heat and light.

2. You can see _____ at night.

3. The _____ moves around Earth.

Day 1

1. Make a list of things you can see in the sky during the day.

_____

2. Make a list of things you can see in the sky at night.

_____

Day 2

1. It is important to never look at the sun. It can hurt your eyes. Draw a picture of something you can wear to protect your eyes on a sunny day.

Day 3

1. Write a sentence telling why you think the moon looks different on different nights.

_____

_____

Day 4

Objects in the Sky

Answer the questions.

Draw a line to match each object to its description.

1.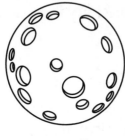

moon

A. These are objects that can be seen in the night sky. They give off light and heat.

2.

stars

B. This is the object in space that moves around Earth. It has craters on its surface.

3.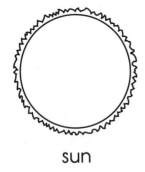

sun

C. Earth moves around this star. It gives Earth light and heat.

4. Count the stars. Write the number. _____

K.RI.4, K.RI.7, K.RF.4, K.W.2, K.L.4, K.CC.A.3

Day and Night

Circle the best word to complete each sentence.

1. Day and night happen because Earth spins, or (slides, rotates).

2. When the sun shines on one part of Earth, it is (daytime, nighttime) there.

1. Circle the picture that is correct.

1. Draw a picture of something you do during the day.

2. Draw a picture of something you do at night.

1. Write a sentence about how your day would be different if it was dark outside instead of light.

_____

_____

Day and Night

Complete each sentence. Use the words from the word bank.

Earth   rotation   sun

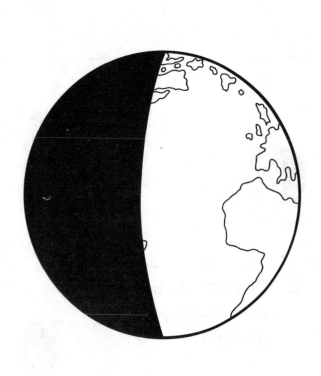 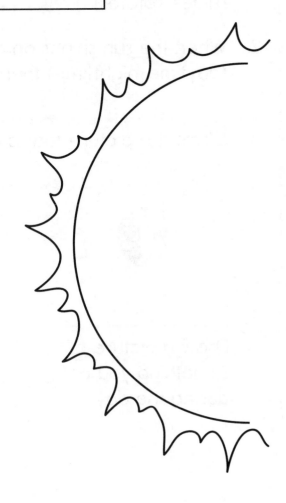

1. Day and night are caused by the _____ of Earth.

2. It is daytime on the side of Earth that faces the

   _____.

3. It is nighttime on the side of _____ that faces away from the sun.

Seasons

Circle the best word to complete each sentence.

1. Spring can be (dry, rainy).

2. It is important for trees and flowers to have water to (grow, move).

Day 1

1. Draw a picture of something you like to do in summer.

Day 2

1. Circle the sentences that are **true**.

   In autumn, leaves grow on trees.

   In autumn, the air is cooler.

   In autumn, the days get longer.

   In autumn, leaves fall from trees.

Day 3

1. Make a list of things you wear in winter to stay warm. How many things did you list in all? _____

   _____   _____

   _____   _____

   _____   _____

   _____   _____

Day 4

Seasons

Circle the name of the season for each picture. Draw an **X** on the item of clothing that does not belong.

1. spring   summer
    autumn   winter

2. spring   summer
    autumn   winter

3. spring   summer
    autumn   winter

4. spring   summer
    autumn   winter

K.RI.1, K.RI.7, K.RF.4, K.L.4, K.CC.A.3

Weather Patterns

1. Circle the weather words.

   rainy   snowy   muddy   windy   sunny

   fuzzy   stormy   cold   sleepy   hot

1. Look at the weather for this week. Draw what you think the weather will be on Saturday.

Sunday	Monday	Tuesday	Wednesday	Thursday	Friday	Saturday

1. Ms. Lee's class tracked the weather for 10 days. It rained on 2 days. It was sunny on the rest of the days. Draw a picture to show how many days were sunny.

1. Write a sentence telling what the weather will be like tomorrow where you live. Use two weather words in your sentence.

_____

_____

Weather Patterns

Look at the calendar. Answer the questions.

1. What will the weather most likely be on the 10th?

   _____

2. On which two days did it rain?

   _____ and _____

3. Draw a picture of what you should wear on the sixth.

4. What type of weather occured the least? _____

5. Write a sentence telling what season you think this is. Give at least one reason.

   _____

   _____

Land, Air, and Water

1. Circle the word that completes the sentence.

_____ is the planet that we live on.

Earth   Moon   Sun

1. Circle the sentence that is **true**.

Earth is made up of land, air, and water.

Earth is made up of rocks, ice, and sunlight.

1. Draw a picture of something that needs air to live.

2. Draw a picture of something that needs water to live.

1. Write three things that land can be used for.

_____

_____

_____

Land, Air, and Water

Answer the questions.

1. Draw a picture of land, air, and water on Earth. Show one living thing and one nonliving thing.

2. Talk with a friend about what might happen if Earth had no water. Write two of the things you think might happen.

_____

_____

Natural or Human Made

1. Natural things are found in nature. Circle the natural objects.

Day 1

1. Human-made things are made by people. Circle the human-made objects.

Day 2

1. Draw a picture of something in the room that is human made.

Day 3

1. Write the name of something made from each natural object.

cotton _____

tree _____

gold _____

Day 4

Natural or Human Made

Answer the questions.

1. Draw a line to match each object to the correct word.

natural

human made

2. Write a sentence telling how many of the objects are natural and how many are human made.

_____

_____

K.RI.7, K.RF.4, K.W.2, K.L.4, K.L.5,
K.CC.A.3, K.CC.B.5

Good Health Habits

1. Draw a picture of something you can do to keep germs away.

1. Order the steps **1** to **4** for washing your hands.

_____ I dry my hands with a towel.

_____ I use soap.

_____ I have dirty hands.

_____ I rinse the soap off with water.

Complete each sentence. Use the words from the word bank.

| fruits | sleep | teeth |

1. Get 10 hours of _____ each night.

2. Brush your _____ twice a day.

3. Eat plenty of _____ and vegetables.

1. Being active is important. Circle the words that name something you can do to be active.

running playing soccer walking

dancing sitting on the couch jogging

cleaning eating a snack

Good Health Habits

Answer the questions.

1. Circle the pictures that show good health habits.

 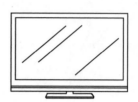

 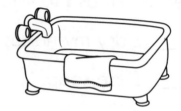

 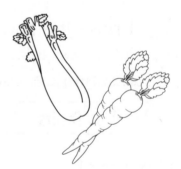

2. How many good health habits did you circle? _____

3. Draw a picture of one more good health habit. Write the total number of good health habits on this page.

    _____

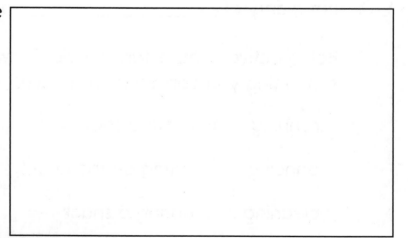

 K.RI.1, K.RI.7, K.RF.4, K.L.4, K.CC.B.4, K.CC.B.5 CD-104811 • © Carson-Dellosa

Dental Hygiene

Draw an **X** on the object in each row that does not belong.

1.

2.

1. Circle the healthy tooth.

   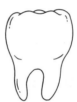

2. Describe each tooth to a friend.

1. Write three things you do to take care of your teeth.

_____

_____

_____

Complete each sentence. Use the words from the word bank.

between	teeth	toothbrush

1. A dentist checks your _____ to make sure they are healthy.

2. Brush your teeth with a _____.

3. Use floss to clean _____ teeth.

Dental Hygiene

Answer the questions.

1. Read each clue. Draw a line to match each clue to its answer.

A. Use this to clean between
your teeth.   toothbrush

B. She makes sure your teeth
are healthy.   floss

C. Use this to clean your
teeth.   toothpaste

D. Put this on your toothbrush
to help take plaque off
your teeth.   water

E. Drink this to keep your
mouth clean.   dentist

2. Casey lost 3 teeth. Pedro lost 7 teeth. How many more teeth did
Pedro lose than Casey?

_____

K.RI.1, K.RI.4, K.RI.7, K.RF.4, K.SL.1, K.L.4, K.OA.A.2

Name_____

Good Nutrition

Circle the best word to complete each sentence.

1. It is important to make (good, bad) food choices.

2. A good meal includes grains, proteins, vegetables, and (candy, fruit).

3. Drink two glasses of (milk, soda) each day.

1. Draw an **X** on each food that is not a good choice.

1. Draw a line to connect each food to its food group.

grains   proteins   vegetables

1. Write three things you will eat or drink today that are good choices.

_____

_____

_____

Good Nutrition

Answer the questions.

1. Draw good food choices for this meal. Be sure to include food from each food group: vegetables, fruits, dairy, proteins, and grains.

2. Liv wants to eat 5 servings of fruits and vegetables today. Liv ate 1 apple and 1 serving of carrots. How many more servings does she need to eat?

_____

Exercise

Complete each sentence. Use the words from the word bank.

| day | exercise | swimming |

1. Your heart works harder when you _____.

2. You should exercise for 1 hour each _____.

3. Walking, dancing, and _____ are all good activities.

1. Write three ways you can exercise with your friends or family.

   _____

   _____

   _____

1. Draw an activity that is exercise.

1. Make a goal to move more. Write a sentence about your goal. Tell an activity you can do. Talk to a friend about your goal.

   _____

   _____

   _____

Exercise

Answer the questions.

1. Circle the sentences that are **true**.

   A child only needs 10 minutes of exercise each day.

   A child should get 60 minutes of exercise each day.

   Walking is good exercise.

   Playing board games is good exercise.

2. Circle the activities you can do to help you exercise for one hour each day. Put a check mark by your favorite activity.

running	sitting
dancing	jogging
reading	cleaning
walking	eating
swimming	writing

Safety

1. Circle the picture that shows how to be safe.

1. Write about how a lit candle can be unsafe. Talk to a friend about what you can do to safely use candles.

_____

_____

_____

1. Riding a bike can be unsafe. Circle the names of objects that can help you be safe.

   seat belt   helmet   gloves   shoes

1. Write a sentence telling why it is important to wear a seat belt when riding in a car.

_____

_____

Safety

Answer the questions.

1. Choose one of the activities below.

    cooking on a stove   cutting paper

    riding in a car   playing outside

2. Draw a comic strip showing how to be safe during the activity.

K.RI.7, K.RF.4, K.W.2, K.SL.2, K.L.4   CD-104811 • © Carson-Dellosa

Scientists

1. A scientist studies objects and spaces. Circle the names of the things a scientist might study.

   water   animals   trees

        movies   weather

1. A zoologist studies animals. Name two animals a zoologist might study.

   _____   _____

2. A geologist studies rocks. Name two places where a geologist might study rocks.

   _____   _____

1. Draw an **X** on each object a scientist probably would not study.

1. Pretend you want to be a scientist when you grow up. Write about what kind of scientist you want to be. Tell what you will study. Talk to a friend about what you wrote.

   _____

   _____

Scientists

Answer the questions.

1. Read the riddles. Draw a line to match each riddle to the scientist's name and the object he or she studies.

A. I study people.
   I help sick people get better.
   I give people shots.

astronaut

B. I study animals.
   I can work at a zoo.
   I visit animal habitats.

doctor

C. I study plants.
   I can work with trees.
   I visit gardens and forests.

botanist

D. I study stars and planets.
   I may go into space.
   I use tools to see faraway planets.

zoologist

2. Choose a scientist above. Write four things the scientist might study.

_____

_____

_____

_____

K.RI.1, K.RI.7, K.RF.4, K.SL.1, K.L.4   CD-104811 • © Carson-Dellosa

Page 9
Day 1: 1. ruler; 2. Check students' drawings.
Day 2: 1. measuring cup; 2. Check students'
drawings. **Day 3:** 1. scale; 2. Check students'
drawings. **Day 4:** 1. stick, hammer;
2. Answers will vary.

Page 10
1. The ruler, measuring cup, and scale
should be circled. 2. Answers will vary.
3. Drawings will vary, but the larger
container (with greater capacity) should
be circled.

Page 11
Day 1: 1. ears; 2. hand; 3. mouth;
Day 2: 1. rose, skunk; 2. balloon, pan;
Day 3: 1. Answers will vary. 2. Answers will
vary. **Day 4:** 1. shoe; 2. cloud; 3. ice-
cream cone

Page 12
1–5. Check students' drawings. The
following should be drawn: red ears,
green skin or hands, blue nose, orange
eyes, and purple mouth.

Page 13
Day 1: 1. The three circles should be
colored. 2. They are circles. **Day 2:** 1. The
car, bike, and truck should be circled.
2. They are forms of transportation.
Day 3: 1. The banana, apple, and grapes
should be circled. 2. They are fruits.
Day 4: 1. The bear, fish, and lizard should be
circled. 2. They are animals.

Page 14
1. house; 2. shirt; 3. suitcase; 4. frame;
5. pencil; 6. Answers will vary.

Page 15
Day 1: 1. on top of; 2. Check students'
drawings. **Day 2:** 1. The picture with the
dot to the left of the rectangle should be
circled. 2. Monitor students' responses.
Day 3: 1–2. Check students' drawings.
Day 4: 1. Answers will vary but should
include a response with a correct
position word.

Page 16
1.
The sun is above the cloud.

2.
The sun is behind the cloud.

3.
The ball is on top of the box.

4.
The ball is below the box.

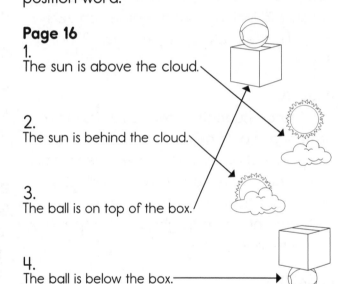

5–7. Check students' drawings.

Page 17
Day 1: 1. zigzag, up and down, straight,
round and round, back and forth, fast and
slow; **Day 2:** 1. Answers will vary. 2. Answers
will vary. **Day 3:** 1. pinwheel, top; 2. Answers
will vary. **Day 4:** 1. up and down; 2. round
and round; 3. straight

Page 18
1. umbrella, arrows up and down; wheel, arrows round and round; swing, arrows back and forth; 2. Monitor students' responses.

Page 19
Day 1: 1. The bird, coyote, and seal should be colored. The seal should be circled. **Day 2:** 1. The ball, fan, and door should be circled. 2. Answers will vary but should include some type of force placed on the ball. **Day 3:** 1. blocks; 2. Answers will vary but should include that the other objects move by themselves. **Day 4:** 1. yes; 2. no; 3. yes

Page 20
1–2. Check students' drawings. 3. The third sentence should be circled. 4. The steeper ramp should be circled. The less steep ramp should be crossed out. Answers will vary but should include the correlation between the ramp height and the speed of the marble.

Page 21
Day 1: 1. force; 2. pull; 3. push; **Day 2:** 1. pull; 2. The wagon should be colored. **Day 3:** 1. push; 2. The boy and swing should be colored. **Day 4:** 1. Answers will vary.

Page 22
1. A. push; B. both push and pull; C. pull; 2. A. 3; B. 3; C. 2; 3. pull; 4. The objects are sorted by the forces needed to move them. 5. Monitor students' responses.

Page 23
Day 1: 1. The umbrella and rain should be circled. **Day 2:** 1. Gravity, down; **Day 3:** 1. Answers will vary but should include that both the leaf and coin are falling to the ground. **Day 4:** 1. Answers will vary.

Page 24
1. force; 2. falling; 3. floating; 4. gravity

Page 25
Day 1: 1. bear; 2. strawberry; **Day 2:** 1. The rectangular pizza should be colored. 2. It is a different shape. **Day 3:** 1. shape; **Day 4:** 1. Check students' drawings.

Page 26
1. Check students' coloring for correct size and shape matches.

Page 27
Day 1: 1. red, brown, green, black; 2. Answers will vary. **Day 2:** 1. The tire and top hat should be colored. 2. The apple and strawberry should be colored. **Day 3:** 1. soft, hard, smooth, rough; 2. Answers will vary. **Day 4:** 1. white, soft; 2. brown, hard; 3. pink, smooth

Page 28
1. pillow, soft; mirror, smooth and/or hard; lemon, yellow and/or rough; table, hard and/or smooth; tree bark, rough and/or hard; crow, black and or soft; 2. Check students' drawings; 7 marbles

Page 29
Day 1: 1. The feather and paper clip should be circled. The brick and table should be crossed out. **Day 2:** 1. 2, 1, 3 **Day 3:** 1. The first two sentences should be circled. **Day 4:** 1–2. Answers will vary.

Page 30
1. light: feather, paper clip, pencil, nail; heavy: house, bear, car, tree; 2. 8 > 6

Page 31
Day 1: 1. Check students' drawings; float; **Day 2:** 1. The paper clip, spoon, and watch should be colored. 2. The ball, boat, and pencil should be colored. **Day 3:** 1. The second sentence should be circled. **Day 4:** 1. Answers will vary. Monitor students' responses.

Page 32
1. A. sink; B. sink; C. float; D. sink; E. float; 2. 2; 3. 3; 5 objects

Page 33
Day 1: 1. paper clip; **Day 2:** 1. table; **Day 3:** 1. wood: stool, skateboard; metal: pan, coins; **Day 4:** 1. Check students' drawings.

Page 34
1. The table, stool, blocks, and pencil should be colored brown. The nail, spoon, pan, wrench, and scissors should be colored blue. 2. metal; 3. 1

Page 35
Day 1: 1. The chair, stool, and table should be circled. 2. wood; **Day 2:** 1. Check students' drawings. **Day 3:** 1. pants, sock, coat, hat; 2. Monitor students' responses. **Day 4:** 1. Answers will vary.

Page 36
1. wood, metal; 2. Answers will vary but should include that a cloth table would not be as strong as a wood or metal table. 3. cloth; 4. Answers will vary but should include that cloth makes clothes flexible, whereas wood and metal would be hard and inflexible. 5. Check students' drawings; 8 tables

Page 37
Day 1: 1. The tearing paper and ice-cream cone should be circled. **Day 2:** 1. stir, cut, break, fold, melt; **Day 3:** 1. Answers will vary but should include that the material and state of matter has stayed the same, yet the shape has changed. **Day 4:** 1. Monitor students' responses. Check students' drawings.

Page 38
1. Tearing paper, slicing bread, boiling water, and the cracked egg should be circled. 2. Answers will vary.

Page 39
Day 1: 1. liquid; 2. shape; 3. poured;
Day 2: 1. The second sentence should be circled. **Day 3:** 1. Check students' drawings.
Day 4: 1. Answers will vary but should include that the water is boiling, turning into a gas, or changing
to steam.

Page 40
1. Check students' drawings. 2. 3; 3. water, liquid; ice, solid; steam, gas

Page 41
Day 1: 1. crab, tree; 2. cupcake, stool;
Day 2: 1–2. Check students' drawings.
Day 3: 1. Answers will vary but may include food, water, air, or shelter. 2. eat, breathe, sleep, drink; **Day 4:** 1. eats; 2. hunts

Page 42
1. 5 nonliving things, 7 living things; 2. red: eat food, breathe, move, drink, grow; blue: make food, grow; Accept different answers as long as students can give reasons to support their choices.

Page 43
Day 1: 1. deer; 2. They live in the ocean.
Day 2: 1. tree; 2. They are flowers.
Day 3: 1. turtle; 2. They are insects.
Day 4: 1. snake; 2. They are birds.

Page 44
1. The penguin, spider, bear, chimpanzee, lizard, and snake should be circled. The pine tree, rose, oak tree, and vine should be crossed out. 2. 6 animals; 3. 4 plants; 4. animals

Page 45
Day 1: The puppy should be circled.
Day 2: 1. Cub, joey, kitten, kid, and bunny should be crossed out. **Day 3:** 1. Answers will vary but may include that a fish lives completely in the water, has scales and fins, and breathes water, while a camel lives on land, breathes air, and has hair.
Day 4: 1. Answers will vary, but drawings should show similar features between mother and baby.

Page 46
1. cow, calf; duck, duckling; cat, kitten; horse, foal; frog, tadpole; chicken, chick; 2. dog; 9 > 6 or 6 < 9

Page 47
Day 1: 1. lion, cat, horse; **Day 2:** 1. penguin, flamingo; **Day 3:** 1. Students should draw scales on the fish and snake.
Day 4: 1. alligator, scales; cardinal, feathers; polar bear, fur

Page 48
1. The chimpanzee, deer, and skunk should be circled. The snake, fish, and lizard should be crossed out. The cardinal, flamingo, and penguin should be colored. 2. 3, 3, 3; 3. equal

Page 49
Day 1: 1. 2; 2. 2; 3. 1; **Day 2:** 1. The first and third sentences should be circled.
Day 3: 1. Drawing should be of a mouth. 2. Drawing should be of a leg or foot.
Day 4: 1. eyes; 2. arm

Page 50

1.

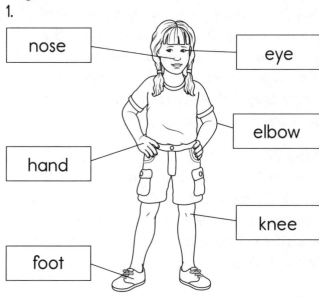

nose

eye

elbow

hand

knee

foot

Page 51

Day 1: 1. help them move; **Day 2:** 1. Check students' drawings. **Day 3:** 1. Both sentences should be circled. **Day 4:** 1. mouse, deer

Page 52

1. tail; 2. wings; 3. ears; 4. fins; 5. hands; 6. Answers will vary.

Page 53

Day 1: 1. stem, flower, leaf, roots; 4; **Day 2:** 1. The second and third sentences should be circled. **Day 3:** 1. The flower, leaf, and roots should be circled. **Day 4:** 1. Check students' drawings.

Page 54

1–4. Check students' coloring.

Page 55

Day 1: 1. flowers, fruit; **Day 2:** 1. The first sentence should be circled. **Day 3:** 1. Check students' drawings. **Day 4:** 1. pumpkin, apple, cucumber

Page 56

1. leaf; 2. roots; 3. flower; 4. stem

Page 57

Day 1: 1. habitat; **Day 2:** 1. owl, bear, fox; **Day 3:** 1. Drawings should show an ocean habitat. **Day 4:** 1. Answers will vary but may include a web.

Page 58

1. ant, anthill; bear, cave; bird, nest; fish, lake; spider, web; 2. 10 animals

Page 59

Day 1: 1. dry; **Day 2:** 1. lizard, mouse; **Day 3:** 1. Drawings will vary but could include a log or rock. 2. Monitor students' responses. **Day 4:** 1. Answers will vary.

Page 60

1. The shark and frog should be crossed out. 2. Check students' drawings; 6 rocks

Page 61

Day 1: 1. cold; 2. polar bear; **Day 2:** 1. penguin; 2. Answers will vary. **Day 3:** 1. Check students' drawings. **Day 4:** 1. Answers will vary but may include that the fur is white.

Page 62

1. Drawings will vary but should include a snowy, bare environment. 2. Check students' drawings; 6 animals

Page 63

Day 1: water; **Day 2:** 1. Check students' drawings. **Day 3:** 1. Monitor students' responses. 2. Answers will vary. **Day 4:** 1. whale, shark, crab, fish, dolphin

Page 64
1. 1; 2. 3; 3. 6; 4. 12

Page 65
Day 1: 1. sun; 2. stars; 3. moon;
Day 2: 1–2. Answers will vary.
Day 3: 1. Check students' drawings.
Day 4: 1. Answers will vary.

Page 66
1. B; 2. A; 3. C; 4. 18

Page 67
Day 1: 1. rotates; 2. daytime; **Day 2:** 1. The picture on the right is correct.
Day 3: 1–2. Check students' drawings.
Day 4: 1. Answers will vary.

Page 68
1. rotation; 2. sun; 3. Earth

Page 69
Day 1: 1. rainy; 2. grow; **Day 2:** 1. Check students' drawings. **Day 3:** 1. The second and fourth sentences should be circled.
Day 4: 1. Answers will vary but may include hat, mittens or gloves, scarf, and coat or jacket.

Page 70
1. winter, swimsuit; 2. summer, snow boots; 3. spring, winter coat; 4. autumn, tank top

Page 71
Day 1: 1. rainy, snowy, windy, sunny, stormy, cold, hot; **Day 2:** 1. Answers will vary but may include rainy. **Day 3:** 1. Students' drawings should show that 8 days were sunny. **Day 4:** 1. Answers will vary but should contain two weather words.

Page 72
1. Answers will vary but may include sunny. 2. first, fourth; 3. Drawings will vary but should include snow gear. 4. rain; 5. Answers will vary but should include winter as the season and snow as a reason.

Page 73
Day 1: Earth; **Day 2:** 1. The first sentence should be circled. **Day 3:** 1–2. Check students' drawings. **Day 4:** Answers will vary.

Page 74
1. Check students' drawings. 2. Monitor students' responses. Answers will vary.

Page 75
Day 1: 1. The watermelon and log should be circled. **Day 2:** 1. The chair and coins should be circled. **Day 3:** 1. Check students' drawings. **Day 4:** 1. Answers will vary.

Page 76
1. natural: bananas, log, rock, tree; human made: table, hat, ring, book; 2. Answers will vary but should include the words *same* or *equal*.

Answer Key

Page 77
Day 1: 1. Check students' drawings.
Day 2: 1. 4, 2, 1, 3; **Day 3:** 1. sleep; 2. teeth;
3. fruits; **Day 4:** 1. running, dancing, walking, playing soccer, jogging, cleaning

Page 78
1. The washing hands, running, bathtub, and vegetables should be circled. 2. 4;
3. Check students' drawings; 5

Page 79
Day 1: 1. hammer; 2. cupcake;
Day 2: 1. The tooth on the right should be circled. 2. Monitor students' responses.
Day 3: 1. Answers will vary but should include using a toothbrush, toothpaste, and floss, and eating healthful foods.
Day 4: 1. teeth; 2. toothbrush; 3. between

Page 80
1. A. floss; B. dentist; C. toothbrush;
D. toothpaste; E. water; 2. 4 teeth

Page 81
Day 1: 1. good; 2. fruit; 3. milk;
Day 2: 1. The candy, cupcake, and ice-cream cone should be crossed out.
Day 3: 1. tuna, protein; bread, grains; celery, vegetables; **Day 4:** 1. Answers will vary.

Page 82
1. Drawings will vary but should include one serving of food from each food group.
2. 3 servings

Page 83
Day 1: 1. exercise; 2. day; 3. swimming;
Day 2: 1. Answers will vary.
Day 3: 1. Check students' drawings.
Day 4: 1. Answers will vary. Monitor students' responses.

Page 84
1. The second and third sentences should be circled. 2. running, dancing, walking, swimming, jogging, cleaning; Students' favorite activities will vary.

Page 85
Day 1: 1. The picture on the left should be circled. **Day 2:** 1. Answers will vary. Monitor students' responses. **Day 3:** 1. helmet, shoes;
Day 4: 1. Answers will vary.

Page 86
1–2. Answers will vary.

Page 87
Day 1: 1. water, animals, trees, weather;
Day 2: 1–2. Answers will vary. **Day 3:** 1. The bowling ball and spoon should be crossed out. **Day 4:** 1. Answers will vary. Monitor students' responses.

Page 88
1. A. X-ray, doctor; B. chimpanzee, zoologist;
C. rose, botanist; D. moon, astronaut;
2. Answers will vary.

CD-104811 • © Carson-Dellosa